Department of the Environment

Department of Transport

Department of Education and Science

Scottish Office

Welsh Office

ENVIRONMENTAL PROTECTION ACT 1990:

CODE OF PRACTICE ON LITTER AND REFUSE

January 1991

LONDON: HMSO

Recycled Paper

© Crown copyright 1991
First published 1991
Second impression 1991

ISBN 0 11 752363 1

CODE OF PRACTICE ON LITTER AND REFUSE

Issued under section 89 of the Environmental Protection Act 1990

INTRODUCTION

Litter has been the subject of much legislation in recent years, brought forward by successive governments. The Control of Pollution Act 1974 (COPA), the Refuse Disposal (Amenity) Act 1978 and the Litter Act 1983, have all sought to bring about a cleaner, tidier local environment.

The 1983 Litter Act provided for a maximum penalty of £400 (now increased to £1,000) for anyone convicted of dropping and leaving litter. Section 22 of COPA apportioned responsibility for cleaning roads in England and Wales between different tiers of local authority. Section 25 of the Local Government and Planning (Scotland) Act 1982 had the same effect in Scotland. Yet, there has until now been no clear and unambiguous requirement for local authorities (or other owners of land to which the public has access, or which is in full view of the public) to keep their land clear of litter and refuse. Part IV of the Environmental Protection Act 1990 ("the Act") makes good this deficiency.

However, the Government recognises that the eradication of litter from our streets and other open spaces will never be achieved simply by attacking the symptoms. Only when there is a universal realisation that dropping litter is fundamentally anti-social and unacceptable will we see an end to a littered local environment.

In order to secure the change in attitude required of many people, the Government has provided grant-in-aid under the Litter Act 1983 to the Tidy Britain Group (which engages in various educational and promotional programmes). Funding has also been provided for the Groundwork Trust and for local environment improvement schemes under the Urban Programme. The work of these bodies continues.

But, despite these measures, it is inevitable that some litter will continue to be dropped and, recognising that litter begets more litter, the Government considers it important to ensure that any litter which is dropped is swiftly and efficiently cleared away. So the Act places a new duty on the Crown, local authorities, designated statutory undertakers and the owners of some other land to keep land to which the public has access clear of litter and refuse, as far as is practicable. (In the case of designated statutory undertakers, the duty may apply additionally to land to which the public does not have access.) The duty also applies to the land of designated educational institutions. The Act requires the Secretary of State, under section 89(7), to issue a Code of Practice to which those under the duty are required to have regard.

Objective

The objective of this Code of Practice is

> "to provide guidance on the discharge of the duties under section 89 by establishing reasonable and generally acceptable standards of cleanliness which those under the duty should be capable of meeting."

It will immediately be apparent that this Code, in its approach to litter clearance, is innovative in at least two ways. Firstly, it attempts, by defining standards of cleanliness which are achievable in different types of location and under differing circumstances, to ensure uniformity of standards across Great Britain.

Secondly, the Code is concerned with *output standards* rather than *input standards* - that is to say, it is concerned with *how clean* land is, rather than *how often it is swept*. Indeed, this Code does not suggest cleaning frequencies at all - it simply defines certain standards which are achievable in different situations. This may mean that an area which all but escapes littering will seldom need to be swept whereas a litter blackspot may need frequent attention. It will be seen, then, that the Code offers considerable scope for local authorities and others to **target** their resources to areas most in need of them, rather than simply sweeping a street because of the dictates of an arbitrary rota. Expressed in its simplest terms: "if it isn't dirty, don't clean it".

Further guidance

The appendix, which is not part of the Code of Practice issued under section 89(7), provides advice on 'best practice' methods which might be adopted to help achieve the described standards (although different practices which achieved the same ends would be perfectly acceptable). Steps which bodies under the duty might take to strengthen public commitment to cleanliness are also suggested in this section.

The Duty

Section 89(1) of the Act places on the Crown and (in England and Wales) county, district and London borough councils, the Common Council of the City of London, and the Council of the Isles of Scilly and (in Scotland) regional councils, district or island councils, joint boards (collectively known as 'principal litter authorities') a duty to ensure that all land in their direct control which is open to the air and to which the public has access is kept clear of litter and refuse, so far as is practicable. In addition, where the duty extends to roads, they must also be kept clean - again, so far as is practicable. (For definitions of the land to which the duty applies see section 86).

Section 86(9) transfers the responsibility for cleaning all roads except motorways (which remain with the Secretary of State) from the highways authorities to the district and borough councils. As regards Scotland, section 86(10) assigns the responsibility for cleaning all roads except motorways (which remains with the relevant roads authority) to the district and islands councils. (For simplicity, the word "road" is used freely throughout the Code and should, where appropriate, be read as "highway or, in Scotland, road". For definitions of the terms "relevant highway" and "relevant road" see sections 86(9) and (10), and 98(5) of the Environmental Protection Act 1990.)

A similar duty is placed on designated statutory undertakers. (Section 98(6) defines "statutory undertaker" as, broadly speaking, bodies authorised by an Act of Parliament to carry on transport-related undertakings. British Rail, London Regional Transport and BAA are just three examples of such a statutory undertaker.) One difference between the duty as it applies to statutory undertakers and the duty applying to principal litter authorities is that the duty on the statutory undertakers might cover some land in the direct control of a statutory undertaker *to which the public has no right of access* (such as railway embankments).

The duty also applies to land in the open air and which is in the direct control of the governing body or local education authority of designated educational institutions (see section 98(2) and (3)).

Similar duties may be imposed by principal litter authorities (other than county councils, regional councils or joint boards) on owners of other land by designating their land as a 'litter control area'. The types of land which may be so designated will be described by the Secretary of State in orders. An important criterion for inclusion in the list of land which might be designated is that it should be land to which the public are entitled or permitted to have access with or without payment. (Examples of the type of land which might be designated are supermarket carparks and privately-owned shopping malls.)

Persons under the duty are required by subsection 89(10) to have regard to the Code of Practice in discharging their duty.

The Act allows the Secretary of State to specify descriptions of animal faeces to be included within the definition of refuse. He may also, by regulation, prescribe particular kinds of things which, if on a road, are to be treated as litter or refuse.

Practicability

The *caveat* in the summary of the duty concerning practicability is very important. It is inevitable that on some occasions circumstances may render it impracticable (if not totally impossible) for the body under the duty to discharge it. It will be for the courts to decide, in all cases brought before them, whether or not it was impracticable for a person under the duty to discharge it but certain circumstances are foreseeable in which the discharge of the duty may be considered by the courts to be impracticable.

For instance, to clear litter or refuse from a railway embankment, or a motorway or other busy road, may entail a restriction of traffic (in the interests of safety both for cleaners and travellers, and to ensure the minimum disruption for road and rail users).

Such restrictions often require detailed planning weeks or even months beforehand, as a consequence of which it may be that litter will have to remain for longer than might otherwise be tolerable. There may be cases where litter or refuse is left as a result of an unforeseeable circumstance or special event. In the first few months after the legislation comes into force, some areas may be affected by the accumulations of many years. Both these circumstances may present particular cleaning difficulties. Similarly, bad weather may make it impossible to clean an area. Moreover, if the weather conditions are exceptionally severe, it may be that resources would have to be diverted to emergency work, making sweeping and cleaning an impracticability. Considerations of safety might also suggest that to clear a very heavily used pedestrian area during peak hours *without risking injury* would be impracticable.

There is also the consideration of what it is reasonable to expect a body under the duty to achieve. For example, whilst the standards in the Code would normally apply throughout the year, it may not be reasonable, or indeed practicable, to expect them to be met on Christmas or New Year's Day.

Enforcement

In the great majority of cases those under the duty will wish to achieve the highest possible standards of cleanliness. However, the Act makes provision for the occasion when a body under the duty may not discharge it adequately. Under section 91 a citizen aggrieved by the presence of litter or refuse on land to which the duty applies may, after giving five days' written notice, apply to the magistrates' court (or, in Scotland, the Sheriff) for a 'litter abatement order' requiring the person under the duty to clear away the litter or refuse from the area which is the subject of the complaint. Failure to comply with a litter abatement order may result in a fine (with additional fines accruing for each day the area remains littered). Any person contemplating enforcement action should not just consider the presence of litter but is advised to consider whether the body in question is complying with the standards in the Code before notifying them, since, under section 91(11) the Code is admissible in evidence in any court proceedings brought under that section.

It may be thought that the requirement on the part of the aggrieved citizen to give the duty body five days' notice before bringing an action, together with the inevitable delay between summons and hearing, will in practice allow a duty body far longer - maybe several months - to deal with accumulations of litter than the period of hours contemplated (for the most part) by this Code (see below). This is not so. Firstly, the courts are specifically empowered (under section 91(12)) to award costs to a complainant where the court is satisfied that *at the time the complaint was made to it* the land was defaced by litter or refuse - even if the land is clean at the time the case comes to court. Secondly, a citizen aggrieved by the persistent failure or wilful refusal of a duty body to discharge its duty would be entitled to apply to the High Court or, in Scotland, to the Court of Session, for Judicial Review of that body's actions.

Similarly, local authorities can act against any other body under the duty which appears to them to be failing to clear land of litter and refuse. Under section 92, if a local

4

authority is satisfied that any land covered by the duty is defaced by litter or refuse, or that defacement is likely to recur, they can serve a 'litter abatement notice' on the person under the duty to keep that land litter-free. The notice may require that the litter or refuse be removed within a specified time and/or that the land must not be allowed to become defaced by litter or refuse again. Non-compliance with a litter abatement notice may result in a fine (with additional fines accruing for each day the area remains littered).

In either case it will be a defence for the person under the duty to show that he has complied with the duty. This Code of Practice is admissible as evidence in court proceedings and if any of its provisions appear relevant to the court in determining any question before it, (eg. whether or not the duty has been discharged) the court shall take account of them.

Cleanliness Standards

This Code of Practice is based on the concept of four standards of cleanliness:

- no litter or refuse, known as grade A;
- predominantly free of litter and refuse apart from small items, known as grade B;
- widespread distribution of litter and refuse with minor accumulations, known as grade C; and
- heavily littered with significant accumulations, known as grade D.

Animal faeces, as prescribed by the Secretary of State in regulations, will not necessarily imply an area has fallen to a specific standard. They may be present where the cleanliness of the area has fallen to grade B, C or to D, and must be considered alongside other litter and refuse.

(Photographs showing examples of the various cleanliness standards in a variety of locations appear in the centre of the Code).

Whilst it is obvious that the first standard of cleanliness (grade A) is the ideal, it is not reasonable to expect that standard to be maintained at all times in all places; technical difficulties may make it impossible to achieve in some circumstances, and it is unlikely to be maintained for long periods in heavily trafficked areas. Grade A should be seen as the standard which a thorough conventional sweeping/litter-picking should achieve in most circumstances - although it is accepted that it may not last for very long. A few items of litter dropped onto a grade A surface will not necessarily be sufficient to degrade that area to grade B.

It will be a matter for the courts to decide how large an area should be considered for the purposes of assessing defacement by litter and refuse, where relevant comparing photographic evidence with the photographic examples in the Code. That will depend on circumstances - whether, for example, a single mound of litter in an otherwise clean street means that the street as a whole has fallen to grade C or D, and therefore the body concerned has failed to discharge its duty.

Zones

The Code has two key principles:

- areas which are habitually more heavily trafficked should have accumulations of litter cleared away more quickly than less heavily trafficked areas; and

- larger accumulations of litter and refuse should be cleared more quickly than smaller accumulations.

The Code therefore divides land types into 11 broad categories of zones according to land usage and volume of traffic. Within the broad descriptions of zones set out below it will be for the local authority or other body under the duty to allocate geographical areas to particular zones (including, where applicable, beaches for the purposes of Category 5 Zone, roads for Category 6 and 7 Zones, educational land for Category 8 Zone, railway embankments for Category 9 and 10 Zones and canal land for Category 11). It is clear that this allocation must be given due publicity, not least to avoid unjustified complaints, although how it chooses to do so will be a matter for the individual body under the duty. Annotated maps in town halls, libraries and other central offices might be appropriate.

Section 95 requires certain local authorities to keep a register on which are recorded details of land which has been designated a 'litter control area' and where street litter control notices issued. Bodies under the duty could use a similar arrangement to publicise their zonings.

In allocating geographical areas to the various zones the duty body will need to use its best judgement, possibly after a period of consultation. It is recognised that there is a level of detail below which it would not be practicable to allocate land to different zones. The duty body will want to avoid, for example, dividing a particular street into 3 different zones simply because it displays characteristics of each of those zones.

Categories

The concepts of standards of cleanliness, practicability and zonings are brought together in this part of the Code which identifies, for different situations, practicable and achievable response times during which the duty body should restore the land in question to a particular condition.

It is stressed that the time periods given below are response times for cleaning an area which has become littered. They do not represent intervals between sweeps, which in many cases could be much longer. Again, 'if it isn't dirty, don't clean it'.

a. General zones

In categories 1 to 4 below, the period from 8pm to 6am is to be discounted for the purpose of assessing compliance with the standards, subject to the proviso in Category 1 Zone below.

It is recognised that on grassed areas, grade A is not always achievable.

Category 1 Zone

So far as is practicable, in town centres, shopping centres, shopping streets, major transport centres (including railway & bus stations and airports), central car parks and other public places where large numbers of people congregate, grade A should be achieved after cleaning. If this falls to grade B, it should be restored to grade A within six hours. If it falls to grade C it should be restored to grade A within three hours and grade D should be restored to grade A within one hour.

If the standard should fall to grade B or below during the period from 8pm to 6am, it should be restored to grade A by 8am.

Category 2 Zone

So far as is practicable, in high density residential areas (containing, for example, terraced houses and flats), land laid out as recreational areas where large numbers of people congregate, and suburban car parks and transport centres, grade A should be achieved after cleaning. If this falls to grade B, it should be restored to grade A within twelve hours. If it falls to grade C should be restored to grade A within six hours, and grade D within three hours.

Category 3 Zone

So far as is practicable, in low density residential areas (containing, for example, detached and semi-detached houses), other public parks, other transport centres and areas of industrial estates grade A should be achieved after cleaning. If this falls to grade C, it should be restored to grade A within twelve hours, and if it falls to grade D it should be restored to grade A within six hours.

Category 4 Zone

So far as is practicable, in all other areas grade A should be achieved within one week of the standard falling to grade C, and 60 hours of the standard falling to grade D.

b. Beaches

In establishing a cleansing standard for beaches careful consideration has been given to the practical difficulties encountered in collecting and removing litter, and the damage to sensitive habitats which may result from such operations.

The duty on any body to clean beaches extends only to the land above (in England and Wales) Mean High Water or (in Scotland) Mean High Water Springs.

Category 5 Zone

Local authorities should identify those beaches in their ownership or control which might reasonably be described as 'amenity beaches'. Any assessment should take into account the level of use of the beach for recreational purposes.

As a minimum standard, all beaches identified by the local authority as amenity beaches should be generally clear of all types of litter and refuse between May and September inclusive. This applies to items or materials originating from discharges directly to the marine environment as well as discards from beach users. The same standard should apply to inland beaches where substantial numbers of bathers or other beach users may congregate.

c.   Roads

Zoning of roads will depend upon their importance within the road network and their environment. For the purpose of this Code, the following hierarchy of roads is adopted:

i.   Motorways

ii.   Strategic routes (all-purpose trunk roads and principal local roads which carry more than 15,000 or 10,000 vehicles per day in urban and rural areas respectively (ie. annual average daily traffic two-way over 24 hour period)).

iii.   All other roads not included in (ii)

Each category can exist within an urban or rural environment. In urban areas category (iii) roads which form an obvious part of the local environment should be subject to the same standards of cleanliness as defined in Zones 1 to 3 above. Urban roads in categories i and ii which, by their nature, are free from pedestrian use (or largely so for non-motorway strategic routes), are clearly used largely to get traffic *through* an area (rather than to service the local area) and are subject to heavy traffic flows, should fall within Zone 6 below. Category (iii) roads not included in Zones 1 to 3, should fall within Zone 7 below.

Although in all cases standards and times relating to clearance of litter or refuse from the roads and related hard-shoulder and verge should be adhered to, as far as is practicable, it is recognised that for reasons of road safety (applying both to those doing the litter clearance and road users generally), and for reasons of avoiding traffic congestion, it will not always be possible to adhere to them. Where this is not possible, the first practicable opportunity should be taken in connection with other maintenance work which takes place to carry out a litter clearance in association with it. These considerations arise particularly in relation to motorways and strategic routes which are subject to continuously heavy traffic flows and where traffic management measures reduce capacity, which may result in severe congestion and delay for users.

In developing a cleansing regime to deal with litter and refuse the duty authority should not forget that it has a duty under section 89 of the Act to keep roads clean. This may include such activities as street sweeping and washing. Care should always be taken to ensure that leaves, debris or litter do not block gulleys, causing flooding.

Category 6 Zone - Motorways and Strategic Routes

On motorways and strategic routes (which may be the responsibility either of the Secretary of State or the local authority), and on associated lay-bys, grade A should be achieved after cleaning of paved areas, and grade B should be achieved after cleaning of verges. If the standard falls to grade C, the area should be restored to grade A (paved areas) or grade B (verges) within four weeks. If the standard falls to grade D, the area should be restored to grade A (paved areas) or grade B (verges) within one week.

In the case of central reservations these time limits shall not apply, but it might be practicable to restore them to grade A and grade B respectively when other work is carried out either on the central reservation itself or in a part of the carriageway immediately adjacent.

Category 7 Zone - Local Roads

For local roads not falling within Zones 1 to 3, and on associated lay-bys, grade A should be achieved after cleaning of paved areas and grade B after cleaning of verges. If the standard falls to grade C, the area should be restored to grade A (paved areas) or grade B (verges) within two weeks. If the standard falls to grade D, the area should be restored to grade A (paved areas) or B (verges) within five days.

d.   Educational institutions

Category 8 Zone - Educational Institutions

The aim when cleaning relevant land of designated educational institutions should be to remove all litter and refuse (grade A).

As a minimum standard during school, college or university terms, grade B should be achieved after cleaning on all relevant land of designated educational institutions. If the standard falls to grade C it should be restored to at least grade B within 24 hours (excluding weekends and half term holidays).

Grade B

Grade D

Grade C

Grade A

Shopping Centres

Residential Streets

Grade B

Grade A

Grade D

Grade C

Grade B

Grade D

Grade A

Grade C

Shopping Centres

Residential Streets

Grade B

Grade D

Grade A

Grade C

Grade B

Grade D

Grade A

Grade C

Parks and Open Spaces

Transport Centres

Grade B

Grade D

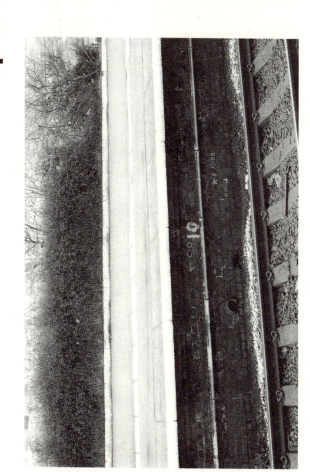

Grade A

Grade C

During other periods, if the land in question is used for a purpose authorised by the governing body or managers of the institution, it shall be restored to grade B within one week of its having fallen to grade C.

Out of term time, where the land in question has fallen to grade C, the governing body or managers shall ensure that it is returned to grade B as soon as is practicable.

e.   Railway embankments

In this section, the references to railway embankments should also be taken to include cuttings, levels and sidings.

In establishing realistic response times and cleansing standards for railway embankments, account has to be taken of access problems, safety of those undertaking clearance work and of rail users and need to avoid disruption to rail services. Due regard should be taken to avoid damage to sensitive habitats.

In areas where the origin of litter or refuse is external to railway activities, clearance arrangements should be on a partnership basis involving the railway undertaking, local authorities and amenity groups, to eliminate blackspots.

Although clearance standards and response times should be adhered to so far as practicable, it is recognised that access, safety and traffic movement may sometimes preclude full adherence. In order to recover the position, the first practicable opportunity should be taken to undertake litter clearance in conjunction with track maintenance work.

The references in Categories 9 and 10 to railway undertakings should be taken to include light railway undertakings.

Category 9 Zone - Railway embankments within 100 metres of station platform ends

Grade B should be attained after clearance. If the standard falls to Grade C, the area should be restored to Grade B within 2 weeks. If the standard falls to Grade D, the area should be restored to Grade B within 5 days.

Category 10 Zone - Railway embankments within urban areas (other than defined in Category 9 Zone)

This category comes into effect from 1 April 1992.

Grade B should be attained after clearance. If the standard falls to Grade C, the area should be restored to Grade B within 6 months. If the standard falls to Grade D, the area should be restored to Grade B within 3 months.

f.   Canal towpaths and embankments

In areas where the origin of litter or refuse is external to the activities of the canal or inland navigation undertaking, clearance arrangements should be on a partnership basis involving the undertaking, local authorities and amenity groups, to eliminate blackspots.

Although clearance standards and response times should be adhered to so far as practicable, it is recognised that access to canal embankments may sometimes preclude full adherence in these areas. (This should not affect adherence to the standards for towpaths.) In order to recover the position, the first practicable opportunity should be taken to undertake litter clearance.

Category 11 Zone - Canal towpaths, to which the public has right of access, in urban areas

On paved areas, Grade A should be achieved after clearance. If the standard falls to Grade C, the area should be restored to Grade A within 2 weeks. If the standard falls to Grade D, the area should be restored to Grade A within 5 days.

On grassed or non-paved areas, Grade B should be achieved after clearance. If the standard falls to Grade C, the area should be restored to Grade B within 4 weeks. If the standard falls to Grade D, the area should be restored to Grade B within 1 week.

APPENDIX

This appendix is not part of the Code. It therefore has no statutory basis.

(For convenience, the term 'duty body' is used in this section to describe any body - local authority or otherwise - under the duty to keep their land clear of litter and refuse, as far as is practicable. Not all advice will be directly relevant to every duty body.)

The Code of Practice describes standards of cleanliness which are achievable in a variety of circumstances and types of location, by those under the duty to keep their land clear of litter and refuse. It does not prescribe the standards nor the way in which the standards should be met as this will vary according to local circumstances and resources. Consequently, this appendix is only intended to be a guide to reviewing policies and practices towards litter. Individual bodies must decide whether this is the right approach for their particular circumstances.

Duty bodies should consider devising programmes which address the problem of litter through (where appropriate) efficient and effective schedules of refuse collection, street cleaning and litter bin servicing, and through public information and education campaigns.

Such programmes might include the following five elements:

- Appraisal
- Action
- Campaigning
- Education
- Enforcement

A: Appraisal

The essential first step in any anti-litter programme is a thorough examination of the facts to assess the nature, extent and cost of litter and littering.

Review of Waste Management Practices

Duty bodies should appraise their current solid waste management practices, cleansing methods, bin provision and servicing, &c... This should provide a quantitative understanding of the relative dimensions of the litter problem faced by the body, thereby providing the framework for identifying the most suitable preventative strategies and cleansing operations.

The appraisal, in combination with measures of public attitudes and awareness to litter, can be used to discover the most effective means of achieving the highest standards of cleanliness within the defined land use zones. This can be accomplished by comparing the data with different litter management and preventative methods.

There may well be scope for targeting the problem areas using the same level of resources previously applied to 'broad spectrum' cleaning operations.

The appraisal should also identify the voluntary groups and commercial and industrial concerns which can help resource the programme in cash or in kind.

B: Action

Duty bodies should consider implementing cleansing/litter removal programmes (prepared following the waste management review) which include such elements as:

- procedures to ensure that grades C and D conditions are dealt with promptly, possibly by the provision of a mobile 'hit squad' to restore the area to grade A condition;

- special arrangements for managing litter and refuse from special categories of land use (eg market stalls where licences could require clearing the area at the end of the day);

- where there is a joint street cleansing and refuse collection contract, consideration should be given to the integration of household refuse collection and the cleansing programme to enable supplementary servicing of litter bins;

- street washing in zone 1, once a week in winter, twice a week in summer, where possible and desirable;

Plans for such programmes, and programme results, should be published to enable the duty bodies and the public to be quite clear about which procedures are being adopted and how effective they are.

A statement should be published by the body explaining the method of quality assurance to be used in monitoring the efficiency of cleansing contractors (applicable whether the cleansing is undertaken by a local authority's own DSO or a private contractor). The method adopted will need to be visual, relate to the four grades of cleanliness described in the Code of Practice and be capable of fairly simple checking.

The duty body should zone its land according to the Code and publish the zones through the local press and by exhibitions in the town hall, library or other public places. (Duty bodies may find useful the Tidy Britain Group's guidelines on zoning land.) Members of the public should be able to ascertain the zone in which they live, work or spend their leisure on request from a named officer of the duty body. There should be published procedures for the receipt and consideration of complaints.

Community Involvement

Members of the public will often be keen to help. The body should encourage participation in a co-ordinated programme of litter abatement activities by providing appropriate support, whether in the form of guidelines, 'tidy codes', publicity material or practical tools to do the clean up job.

Design against Litter

In all new developments and in repair and maintenance works, particularly in road and paved surfaces, the designs should have regard to the subsequent cleaning and maintenance practices. Guidelines should be prepared indicating location, style and capacity of the litter receptacles, and provision made for waste storage facilities.

C: Campaigning

An effective public information campaign is essential. The community needs to know the size of the litter problem, what steps are being taken to solve it, how they can be involved in the cleaning up operation and where success has been achieved.

The expertise in public relations of the duty body should be employed. Materials, such as posters, stickers, badges, T-shirts, films, videos, etc, could be produced to stimulate interest and to publicise the campaign. The local media should be encouraged to make the campaign their own and publish and broadcast news about the problem, the action and the successes. Competitions such as 'best kept street/village/ industrial estate' should be encouraged as they reinforce success. Sponsorship for these can be sought from the public and private sectors.

D: Education

Education has an essential role in awakening the interest of children and young people in all aspects of their environment and in cultivating in them informed concern about environmental problems, including the problem of litter in our streets and public places, together with a sense of personal responsibility for the quality of the environment.

Environmental education is an important theme which runs through the National Curriculum for children aged 5 to 16 in local education authority and grant maintained schools in England and Wales; and in parallel national curricular initiatives in Scotland. It will be taught in the context of many of the foundation subjects, including science, geography, and technology. Through these studies, as well as through personal and social education, which will remain a feature of the wider curriculum, teachers and schools will have a vital part to play in shaping positive attitudes to tidiness and litter abatement.

E: Enforcement of the offence of leaving litter (section 87)

The local police and, in England and Wales, magistrates' clerks should be invited to join in any discussions about enforcement of the legislation. Local authorities could include in any anti-litter publicity information about:

i.   the operation of the fixed penalty scheme for leaving litter (section 88), for example

        -    where it will operate, and
        -    which officers will enforce it.

ii.   schemes (either voluntary or statutory under section 99 of the Act) to deal with abandoned shopping or luggage trolleys.

Care must be taken not to place weight on enforcement of the offence until the local authority waste management services are fully effective and the information and education programme has had time to create awareness.

Local authorities, and other bodies with byelaw making powers, should strongly consider applying the full range of dog byelaws, as appropriate, on their land. Such byelaws include powers to, in certain areas, ban dogs, require dogs to be kept on a lead, require the owner to clear up after his dog ('poop-scoop' byelaw) and make it an offence for the owner to allow his dog to foul.

MISCELLANEOUS ADVICE

Training

Duty bodies should have regard for the training of their staff in litter abatement practices. Litter can be a hazard in the work place causing accidents and fires if it is not handled correctly. The individual's role in waste management is important and it should be part of the organisation's own policy of litter control and waste management to emphasize it. If this policy is to be successful, it must be the responsibility of a member of the management team and must be written into his or her job description.

Complaints

As explained in the introduction, the Act gives aggrieved citizens the right to seek redress in the courts if a duty body fails to keep its relevant land clear of litter and refuse (and, in the case of relevant roads, clean). However, before applying to the courts the aggrieved citizen must give to the duty body five days' written notice of his intention. It is in the interests of all parties, therefore, for the duty body to publicise the best means for an aggrieved citizen to make his complaint known.

Local authorities who have the responsibility for cleaning roads will notice that, in certain circumstances, the Code provides for land to be restored to grade A within two or four weeks - far longer than the statutory five day notification period. It is conceivable, therefore, that the situation may occur where a person, aggrieved by the littered state of a road in zone 6 or 7, may give notice of intended court action without being aware that the Code suggests longer response times (for example, in category zone 7, if the standard falls to grade C, it should be restored to grade A (paved areas) or grade B (verges) within two weeks). To pre-empt court action local authorities may wish to have a standard response immediately available to send to complainants in such cases, explaining this position. This advice also applies, in certain circumstances, to statutory undertakers and educational institutions.

Litter bins

The size and siting of litter bins is a matter which can only be decided with reference to conditions prevailing in each area, although empirical research clearly shows that bus stops, outdoor seating areas and the environs of take-away restaurants need plenty of well-serviced bins.

The frequency of litter bin placement on footways is, likewise, a matter for local determination, but research suggests that bins sited at frequent intervals may be appropriate in particularly busy thoroughfares, such as high streets in zone 1. Authorities should also consider the provision of bins in other zones, particularly parks, car parks and beach areas, and the temporary provision of bins at special events. When siting bins the needs of the disabled and visually handicapped should be remembered.

The Tidy Britain Group has more detailed advice on the size and siting of bins. For further information contact the Group at The Pier, Wigan WN3 4EX (or in Scotland, contact Keep Scotland Beautiful at Old County Chambers, Cathedral Square, Dunblane FK15 0AQ).

Fly-tipping

Any statutory undertaker, or any other duty body, which suffers fly-tipping on their land should take steps, once accumulations of refuse have been initially cleared, to prevent recurrence. These steps may include the following:

-       preventative measures such as higher fences along boundaries, or maintaining existing structures.

-       identifying those responsible for fly-tipping and persuading them to clear up, followed by prosecution (or the threat of it in the first instance) where co-operation is not forthcoming.

-   setting up liaison machinery between the duty body, the local authority (as appropriate) and local residents and voluntary groups. This could tie in with education on the value of a tidy environment, together with information on alternative amenity tip or local authority collection facilities in the particular area.

Grass cutting

Where grass cutting takes place on relevant land or a relevant road, duty bodies should take steps, as far as is practicable, to co-ordinate that activity with litter clearance, whether or not they are responsible for cutting the grass in question. It may be appropriate to arrange for litter clearance before cutting where the grass is kept short, such as on playing fields, whereas on grassed areas which are not so heavily maintained, such as verges, it may be more appropriate to arrange litter clearance after cutting when the litter and refuse will be more exposed.

Duty bodies are encouraged to make arrangements to remove litter and refuse from any hedge or fence adjoining the relevant land or road, where it is accessible and visible from such land or road.

In general, when clearing litter and refuse from vegetation, care should be taken to avoid damaging sensitive wildlife habitats. If in any doubt as to the value of any area for wildlife, independent advice from an appropriate nature conservation body should be sought before litter clearance is undertaken.

Printed in the UK for HMSO
Dd 0293045 2/91 C570 563746 12521